Factories that Build Factories

Eric H. Anderson

Seattle   Discovery Institute Press   2024

I0075167

Description

Evolutionists acknowledge that without a self-replicating entity, the Darwinian process has nothing to work with. So how could mindless chemicals have built the first self-replicating entity to kickstart Darwinian evolution? As design theorist Eric Anderson explains, evolutionists suggest that something simple—like a self-replicating molecule—kickstarted the origin of life on Earth. But is that idea realistic? As it turns out, engineers are trying to build a self-replicating machine, and while they are nowhere close to succeeding, their efforts reveal that even the simplest self-replicating system must be unimaginably sophisticated. Mindless processes aren't up to the task. What's required is a most intelligent designer.

Library Cataloging Data

Factories that Build Factories by Eric H. Anderson

38 pages, 6 x 9 in.

ISBN-13 Paperback: 978-1-63712-055-2, Kindle: 978-1-63712-056-9, EPub: 978-1-63712-057-6

BISAC: SCI017000 SCIENCE / Life Sciences / Cell Biology

BISAC: SCI027000 SCIENCE / Life Sciences / Evolution

BISAC: SCI075000 SCIENCE / Philosophy & Social Aspects

Publisher Information

Discovery Institute Press, 208 Columbia Street, Seattle, WA 98104

Internet: https://discovery.press/

Published in the United States of America on acid-free paper.

First Edition, First Printing, August 2024.

CONTENTS

Factories That Build Factories

Eric H. Anderson

Nobel Prize recipient and Harvard origin-of-life researcher Jack Szostak once remarked, "In my lab, we're interested in the transition from chemistry to early biology on the early earth.... You want something that can grow and divide and, most importantly, exhibit Darwinian evolution."[1]

Another noted origin-of-life researcher, Gerald F. Joyce, says much the same thing. When asked about the idea that chemicals might have come together on the early Earth to form something that could copy itself, Joyce responded, "That's what we and others are interested in because that's sort of, you know, the tipping point between chemistry before and biology after."[2]

Self-replication, then, is not just one more in a long list of problems to be solved for the origin of life. As far as many of the leading origin-of-life researchers are concerned, discovering the pathway to a self-replicating entity is the central challenge, the Holy Grail. Figure out how to get that from purely natural processes, and the hope is that everything else will take care of itself.

But it's a Grail that continues to elude the research community, despite the brash claims occasionally made to the contrary.

Dawkins's Miracle Molecule

A FEW years ago I happened to turn on my car radio and caught the end of a lecture segment on public radio. Evolutionary biologist and prominent atheist Richard Dawkins was the guest. Dawkins held the position of Professor for the Public Understanding of Science at Oxford University for more than a decade, and one of the questions posed to him made me quickly reach over and turn up the volume.

"How close are we to understanding the origin of life?" the moderator asked.

I half-expected Dawkins to acknowledge the many difficulties with abiogenesis, to admit that this was a huge open question, and to confess that we don't yet have any good abiogenesis scenarios, while claiming, as so many proponents of evolution do, that the origin of life is a separate question from biological evolution. That is, I thought he might concede the many widely acknowledged difficulties still facing the origin of life but try to contain the damage for the materialistic outlook by emphasizing that at least things were well in hand for evolutionary theory after the origin of the first life.

To my surprise, Dawkins responded rather glibly that we have a pretty good idea how life started. Yes, there are some challenges, he acknowledged, but we know what happened in broad strokes and at this point, he implied, we are basically filling in the details.

Having studied the origin of life at length and being aware of the many and acute problems with abiogenesis theories, it struck me as more than a little irresponsible for someone wearing the title of "Professor for the Public Understanding of Science" to claim in a public venue to thousands of listeners that we have a pretty good idea how life started.

Why would Dawkins make a statement like that? Was he purposely misinforming listeners about the current state of the science, or was he unaware of the many problems with abiogenesis? Did he really believe what he was saying?

As I analyzed the question further in the coming days, I realized that Dawkins's thinking likely stems from the notion that the origin of life—at least the initial starting event—was a relatively simple event. Not necessarily a common event or an easily repeatable event, mind you, but a relatively simple one.

In his book *The Selfish Gene*, Dawkins paints a picture remarkably similar to Darwin's statement in his 1871 letter to Joseph Hooker (quoted in the previous mini-book). "Nowadays large organic molecules would not last long enough to be noticed: they would be quickly absorbed and broken down by bacteria or other living creatures," Dawkins writes. "But bacteria and the rest of us are late-comers, and in those days [on the early Earth] large organic molecules could drift unmolested[3] through the thickening broth."[4]

With this assumed backdrop of early Earth conditions, Dawkins goes on to suggest the first key step in the origin of life: "At some point a particularly remarkable molecule was formed by accident. We will call it the *Replicator*. It may not necessarily have been the biggest or the most complex molecule around, but it had the extraordinary property of being able to create copies of itself."[5]

This hypothetical self-replicating molecule is crucial to the materialist creation story, and on two counts. First, getting a complete organism to arise by chance is, as is widely acknowledged, too unlikely and never could have occurred. So something simpler, something that had a much greater likelihood of arising by pure chance, something like a simple self-replicating molecule, had to kick-start the process. Second, once this self-replicating molecule came on the scene, then Darwinian evolution could kick in, bringing the impressive power of random mutations and natural selection to eventually transform our simple self-replicating molecule into an actual organism.

At least that is how the story goes.

This "particularly remarkable molecule," Dawkins suggests, is easy to imagine, and the remainder of his description of this extraordinary entity consists of a simple, though chemically unrealistic, thought experiment about how such a fascinating molecule might work, making copies of itself, "competing" with other molecules in the watery environment, and so on.

Origin-of-life researchers, to their credit, haven't been satisfied with thought experiments alone. There has been a great deal of effort expended over the past couple of decades trying to create a self-replicating molecule in the lab, and then to apply the lessons learned to the question of the origin of life. Some good work has been done and some interesting results occasionally published, but nobody has been able to create such a molecule.

To be sure, there have been several papers published and news stories released proclaiming that researchers have created this or that self-replicating molecule, but these claims invariably turn out to be misleading. If anyone has actually discovered or created a self-replicating molecule, they are keeping it a very good secret.

This failure to produce such a molecule, keep in mind, is despite decades of research and lavish financial expenditure. The reason for the failure is not for lack of time, effort, and funding. No, the reason is much more fundamental.

The Blob Has a Secret

THERE WAS a sense in Darwin's day that microorganisms were rather simple, each one little more than a tiny "blob of protoplasm."[6] Darwin viewed the organism as a flexible conglomeration of these simple cells. Through no fault of their own, he and his contemporaries of the time knew nothing of genetic information processing, signaling, and feedbacks, nothing of cellular machinery, integrated systems, complex coordination of molecular parts, or the many other requirements for even the simplest working cell.

In *The Origin of Species* Darwin described organisms as "plastic."[7] He wasn't referring to the material used to make children's toys today, but rather to the idea that organisms were flexible and could, he was convinced, be readily shaped and molded by natural selection to essentially any form. From this viewpoint it followed that adding more cells or making changes to the organism should also be a relatively simple process.

However, with the accumulating knowledge of cellular structures in the late 1800s, the discoveries of cellular systems and proteins and metabolic pathways, the unraveling of DNA's structure in the 1950s, and the subsequent discoveries up to the present that continue to uncover new depths of biological complexity, it became ever clearer that cells are anything but simple, and that even the humblest organism is complex beyond anything previously imagined. Not just complex. Complex and coordinated, with a 4-bit digital code, information storage, retrieval and translation mechanisms, error-correction algorithms, functionally integrated systems, and molecular machines—marvels of nanotechnology that put to shame anything humans have yet created.

As a result of these discoveries it became increasingly clear that no organism, even a relatively simple single-celled organism, could arise all at once on the early Earth by chance.

But if life couldn't arise by chance as a single event, perhaps a series of events could do the trick. Perhaps if the problem were broken down into simpler steps then it might be possible?

With that thought firmly in mind, abiogenesis proponents busily churned out hypothesis after hypothesis that might help the process along, simpler steps that could perhaps lead to something more. In other words, rather than attempting to explain how a simple single-celled organism could arise by chance, many origin-of-life researchers focused on identifying what the initial capacity or characteristic of the earliest stage of life likely was. Information storage and processing seem to be central to all life forms. So perhaps life started with DNA. Other researchers noted that life needed a way to obtain and use energy to run cellular processes. So perhaps the key to the origin of life was a primitive form of metabolism. Still other researchers focused on the fact that organisms needed a way to protect themselves from the surrounding environment and protect the tender early chemical reactions from interference. So perhaps life had to start with a protective shell or bubble, some kind of early cell membrane.

One of the more promising ideas emerged from the discovery that some RNA molecules could act as enzymes, helping catalyze chemical reactions in the cell. Since RNA could also store information, similar to DNA, it seemed ideally suited to perform not just one, but two roles in the abiogenesis story. So perhaps, it was suggested, life started as an RNA molecule.

These and other ideas continue to be developed today, with regular press releases and articles gracing the pages of popular news sites and prominent scientific journals. Without minimizing the importance of any particular avenue of research, it is nevertheless safe to say that the most prominent view today among origin-of-life researchers is essentially the same that Dawkins outlined in *The Selfish Gene*—namely, life started with a molecule, some kind of self-replicating molecule.

Darwinian Evolution All the Way Down

THE CURRENT evolutionary view of life, although much more attuned to the complexity and information-rich properties of organisms than was the science of Darwin's time, is still very much dependent on the same two assumptions Darwin made so long ago: *First, if we start with a simple entity it will eventually undergo variations sufficient in both kind and quantity to turn it into another organism; indeed, to eventually turn it into everything we see in biology today. Second, organisms are very flexible in their makeup, able to undergo innumerable changes over time and incorporate those changes into their makeup without breaking down or missing any steps in the long chain of required changes.*

Based on these two fundamental assumptions, the thinking among current origin-of-life researchers is that if we can just get a self-replicating molecule on the early Earth, then the Darwinian process of mutations and natural selection can take over and, eventually, produce the first living organism. Then that first living organism will of course go on to produce everything we see around us in nature…

Don't misunderstand. There is serious doubt (as we will explore later in this mini-book and in others in this series) whether the evolutionary mechanism can actually transform a self-replicating molecule into a living organism and produce the kinds of systems we see in living beings. But the vision of natural selection having near-mystical powers of creation has taken such hold on the evolutionary imagination that many researchers seem to believe that with the magic wand of natural selection, "all things are possible." Stated another way, it is not that there is actually any good evidence that a self-replicating molecule can give rise to complex life; it is just that once natural selection kicks in, the idea becomes more believable to many people.

As a result, many prominent origin-of-life researchers today view the origin of life as essentially a problem of getting a single self-replicating molecule. Once a self-replicating molecule appears on the hazy scene of the early Earth, the power of Darwinian evolution is thought to take over and then… anything goes.

Thus does an article discussing origin-of-life research by David Horning and Gerald Joyce observe, "Research into the origin of life takes for granted that the first living thing was much simpler than any existing life… At some point, one molecule acquired the ability to replicate itself from chemicals found on the primordial Earth. Once that took place, Darwinian evolution could take over."[8]

With that assumption firmly in place, the self-replicating molecule takes on the central role in the dramatic history of life on Earth, and self-replication becomes the key to the entire abiogenesis story.

Most criticisms of abiogenesis over the years have focused on the reducing atmosphere, energy sources, the difficulty of forming life-essential polymers in the primordial soup, the existence of the necessary nucleotides or amino acids at the right place and time, the rise of coding and information-rich molecules, the astronomical odds against getting molecules joined up in the right order, and so on.

However, much less time has been spent on the issue of self-replication. Indeed, even most critics of abiogenesis largely ignore the issue or appear to accept unreflectively the idea that self-replication may indeed be an early step on the path to the first life.

Evolutionary proponents, for their part, hesitated to take on the challenge of abiogenesis for many decades. Indeed, for debating purposes many refused to grapple with it, arguing that the origin of life is a problem that stands completely apart from evolution and need not be addressed by evolutionary proponents because evolution only starts once there is a living organism. But the formidable challenges with abiogenesis have forced its proponents to fall back on what they view as the most powerful force of creativity: Darwinian evolution in the form of natural selection acting on variations arising within self-reproducing entities thought to exist before the first living cell. The hope is that in this way Darwinian evolution can solve the difficult challenge of getting that initial living organism in the first place, a challenge that includes creating metabolism, cellular systems, and pages and pages of precisely coded, high-functioning digital information.

So, rather than being irrelevant to the origin-of-life story, Darwinian evolution is now viewed as central to it—a last-ditch effort to hold together the many rapidly unraveling threads of the abiogenesis tapestry. This marks a significant shift in the rhetorical underpinnings of the materialist creation story.

It is true that earlier researchers such as Oparin, Haldane, and other abiogenesis proponents had at times described the origin-of-life process as some kind of chemical "evolution." Yet any connection between such a chemical evolutionary process and Darwinian evolution long remained tenuous at best. Indeed, many scientists and theorists over the years explicitly rejected the application of Darwinian evolution to the time period before the existence of the first living organism, drawing a bright line between the problem of the origin of life and the evolution of life following its origin.[9]

This is no longer the case. Under the current paradigm, abiogenesis begins with undirected chemical reactions, leading by chance to a simple self-replicator, such as a self-replicating molecule of some kind. Abiogenesis proponents imagine natural selection stepping in at that point to help the tender molecule acquire additional traits, eventually leading to the first living organism and then from there to the full complexity and diversity of life we see today.

So, rather than having a living organism and then endowing that organism with an *additional* ability of self-replication, the materialistic abiogenesis story makes self-replication the first ability. Self-replication thus becomes the initial formative characteristic of the ancestral entity of all life, the characteristic from which all other subsequent features and powers arise.

Three Indispensables… Actually, Four

WHAT WOULD it take for a chemical unit on the early Earth to achieve the ability to self-replicate? That's no small challenge. Some have pointed to crystals, autocatalytic reactions, and even RNA ligase molecules.[10] But despite interesting chemical and structural properties, none of these actually self-replicates.

A number of researchers have considered what might be required of even the simplest truly self-replicating chemical entity. Here are three core requirements: (1) It has to be able to form under natural conditions, without help from a lab technician. (2) It has to be able to make copies of itself by locating and ordering specific atoms or small molecules that would be available in the environment, not by simply catalyzing a reaction between carefully designed and previously prepared sections of itself, as in so many of the lab experiments carried out in recent years. (3) It has to be stable enough to exist in real-world conditions on the early Earth—the "primordial soup"—without breaking down too quickly and without getting bogged down by interfering cross-reactions.

Additionally, if the molecule is to jump-start a Darwinian process of mutation and natural selection, it must have the capacity to mutate while somehow retaining the ability to faithfully replicate its now-mutated self.

Based on extensive research performed by many origin-of-life researchers over the past couple of decades, there are good reasons to conclude that a single molecule alone could not manage all this.[11] Note, too, that the above is only a summary list of requirements. A comprehensive list would fill many pages.[12]

The challenge in origin-of-life studies is to self-replicate a self-replicator, which can in turn self-replicate a self-replicator, *ad infinitum*. The equivalent would be a computer that can self-replicate itself, the copies of which could self-replicate themselves, on and on. And indeed, not just computers but fully functional robots.

Some might be tempted to point to a computer virus or other computer program that can copy itself, but such programs are not self-replicating in any substantive sense. The software program only exists and runs on a carefully designed and functional piece of hardware that is certainly not replicated in the process. Further, there is generally an operating system, as well as several additional layers of software in the form of drivers, compilers, interfaces, middleware, and so on. The most that can be said for so-called "self-replicating" computer programs is this: a carefully designed and functionally integrated system of hardware and software can reproduce a copy of *a portion* of the software on the machine. Interesting, yes, but essentially irrelevant to the problem at hand.

True self-replication is a more onerous task. Once we consider the task of actually building a self-replicating machine that can exist in the real world in physical, three-dimensional space, the challenges become a bit more evident.

A 3D Printer That Prints… 3D Printers?

To HELP us understand what is involved in self-replication, let's set aside for a moment the dizzying complexity of the living cell and consider what would be involved in building the simplest possible self-replicating machine with our existing technology.

The ability to engineer a self-replicating machine has been the topic of much discussion in science fiction stories and movies, ranging from the large and powerful Terminator-style robots to small but deadly nanites. It remains that—science fiction. But will it always be so?

There are now 3D printers with the ability to create some of their own parts in three-dimensional space. This has, for the first time in human history, allowed us to begin dreaming in concrete terms of taking the first steps on the long and still baffling road to creating a self-replicating machine. Is such a feat even possible? It sounds fantastic, even crazy, but the fantastical idea is driving some fascinating research. And what we are learning has direct implications for research on what would be minimally required for the first self-reproducing entity in the history of life.

First, a bit of background on 3D printing for those new to the field.

I became interested in 3D printing years ago and have followed the development of the industry off and on ever since. Recently the technology has become cheap enough that 3D printers have moved into the world of the hobbyist and the technology enthusiast. Popular consumer-level printers include MakerBot, FlashForge, Creality3D, and many others.

At the time I am writing this, 3D printers range from personal machines costing a few hundred dollars that produce rough prints in a single material to high-end professional printers costing many thousands of dollars boasting sub-50-micron resolution and printing in multiple materials. Several different kinds of 3D printing technologies also ex-

Figure 1. 3D printed stand designed by the author on an early extrusion printer.

ist, from material extrusion (the most common technology for consumer and prosumer printers, and the one that you have probably seen many times), to powder bed fusion, photo-polymerization, ultrasonic additive, material jetting, electron beam melting, and more.

Several years ago our local library received a grant to educate library patrons on the technology of 3D printers by allowing patrons to reserve print time, so I took advantage of the opportunity to design and print a simple stand for one of my 5x5x5 cubes (basically a Rubik's cube on steroids).

This is an exciting and explosive technology that promises to transform the landscape of design and prototyping activities, and even some manufacturing processes. There are a lot of daring engineers and visionaries in the field, but perhaps among the most daring are those involved in the RepRap Project, an open-source venture that seeks to create a self-replicating 3D printer.[13] Many people have been involved in this project and have done tremendous work in moving it forward, with significant strides made. Many of the parts for a RepRap printer can be printed on the printer itself to reasonable tolerances, enabling a hobbyist to use those parts in the construction of a new printer.

However, as is often the case with groundbreaking new technologies, the excitement around potential future breakthroughs tends to intrude into assessments of on-the-ground realities. For example, the RepRap website touts the machine as "humanity's first general-purpose self-replicating manufacturing machine." In an allusion to evolutionary thinking, and ignoring the tremendous amount of design and engineering involved in producing RepRap, one of the early RepRap printers was even named "Darwin."

Another interesting printer, the Kickstarter-backed BI V2.0,[14] received breathless attention in late 2013, with a myriad of headlines touting "The World's First Self-Replicating 3D Printer!" This isn't just sloppy newswire enthusiasm; even the official project website touted BI V2.0 as "a self-replicating, high precision 3D Printer."[15]

If one is the trusting sort, it's tempting to look at projects like RepRap or BI V2.0 and think, "Wow! We are almost there. We've nearly created a self-replicating machine!" But a closer look is warranted. Neither RepRap nor BI V2.0 are self-replicating.

Not in theory or in practice.

Not even close.

Not even in the ballpark.

Don't get me wrong. I love this technology. I've followed RepRap closely and consider it a fantastic idea and an excellent open-source project. I even considered contributing funds to the BI V2.0 Kickstarter project when it first came out. But despite impressive efforts, a human-designed self-replicating machine remains a long way off.

Finding the Ballpark

MANY REPRAP users have shared online photographs of the many parts of the RepRap printer that can be printed on the printer itself in a hard plastic material, like ABS plastic.[16] An impressive number of parts can be printed, to be sure. Yet even a cursory analysis of the printing process reveals that the printer is not actually able to print any of those parts by itself.

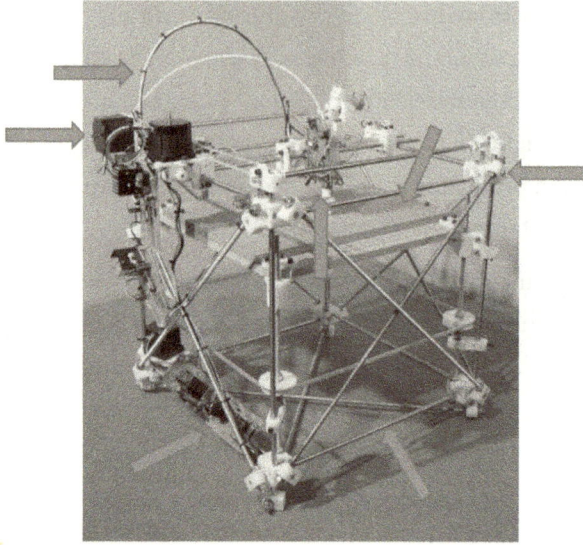

Figure 2. RepRap printer with arrows pointing to just a few of
the many parts that cannot be printed on the printer.

The printer must first be set up and programmed with the right parameters, and it must be fed the material for extrusion. Even after parts are printed, they must be carefully removed from the print bed by the user, inspected for quality and, in many cases, cleaned up or sanded before they are ready as usable parts. So the impressive number of parts that can be printed doesn't tell the whole story.

What's more, many of the printer's other parts cannot be printed by the printer, with or without help. These include structural support rods, screws, copper wire, rubber drive belts, a precision stainless steel extrusion nozzle, a print bed, a heating element, clamps and ties, and a filament feeder.

More daunting still, the printer requires a circuit board, an SD card reader, cabling, control switches, electric motors, and more in order to function.

The printer is not even close to being able to produce all those parts.

And the situation is even worse than all those problems would suggest.

An electric motor by itself is a precision machine that requires multiple parts, manufactured to tight tolerances and assembled in just the right way in order to function. Even a simple electric motor includes a housing, shaft, rotor, stator, terminals, magnets, copper windings, some kind of lubricant and/or bearings, and wiring connections.

More daunting still, a printed circuit board is a marvel of modern engineering and includes thermoset resin laminate materials, a battery, resistors, transistors, capacitors, inductors, diodes, switches, and more— all manufactured to precise tolerances.

Finally, for our printer to be truly self-replicating, it must be able to work in multiple materials. It is one thing to print parts in plastic and to have the right nozzle and heating element to do that. It is quite another to print parts in multiple materials. If our printer were truly self-replicating, it would be able to work not only in plastic, but also in aluminum, stainless steel, copper, rubber, silver, gold, tin, and fiberglass. It also would need the ability to safely handle zinc chloride, hydrochloric acid, resin laminates, and etching chemicals.

This is only a partial list of missing requirements.

Now, it's true that savvy engineers and designers often can find ways to substitute different materials and in different ways, so it's possible to find a few workarounds that would trim down the total number of different parts and materials. To that end, some enterprising individuals have created offshoots of RepRap that use plastic in place of the steel support rods and the rubber belts. This is clever design work. But remember, these workarounds are limited in their application, the description above is just a very partial sketch, and a detailed analysis would reveal many more parts and materials required for a truly self-replicating printer.

The point here is not to dog on RepRap. As I said, I'm a fan. The point is to provide the reader with a sense of the challenge involved and the scale of the problem. Whichever printer we are analyzing and whichever parts list we want to focus on, we find that creating even a relatively modest machine that can produce all of its own parts is an astounding engineering challenge—a challenge far beyond anything our best engineers and brightest minds can presently accomplish.

Putting It All Together

IF WE are hoping to ever build a self-replicating machine, there is another aspect of the process we will have to master that goes beyond replicating the parts. Even if our printer had the remarkable ability to print multiple materials at the sub-micron level, and print every single part used in the construction of the printer—something which at this stage is but a distant dream—it would still not be able to assemble itself. To be truly self-replicating, it would also need to be able to assemble those parts in actual, physical, three-dimensional space.

To do that, the printer would need to have carefully controlled and sophisticated robotic assembly systems. For example, it would need an assembly arm to pick up the printed pieces, analyze them for completeness and quality, rotate them into the right position, and place them in the correct location. In reality, this would require multiple assembly arms and mechanisms.

Also, as soon as we introduce a new assembly arm and its many attendant parts into the mix, we then have an entire additional set of machine parts that in turn have to be incorporated into our printer design, printed, and assembled. It also would require additional computer software. And that software, that digital information, would be far from simple. Indeed, the entire printer would need to be radically re-engineered to account for these additional parts and mechanisms.

Worse, every time we include a new part or an additional mechanism to assist with this challenging self-replication process, that new part or mechanism also must be replicated, requiring additional instruction sets, perhaps a reworking of the machine's physical layout, and additional information related to this new part or mechanism: how it is to be constructed, how it is to be assembled, how it is to function.

Indeed, every single time we add a significant new part, or in the vernacular of the materialistic evolution story, every time the nascent organism evolves a new meaningful function, that new function requires not only a careful integration into the whole, but the instruction set to implement and reproduce the parts necessary for that new function.

Further, it is unclear how the printer could even accomplish the assembly without additional significant re-engineering. Remember, the printer occupies a physical three-dimensional space. The best it can do is assemble a copy right next to itself, with the far side of the copy some 12 to 24 inches away. Thus, any assembly mechanism would have to be able to reach outside of the box—outside of itself—in order to reproduce itself. This suggests another challenge.

In-House Engineering

A 3D printer assembling a copy outside of itself might work on a clean tabletop with no other interference, but such an approach is unfortunately completely unworkable in the fluid and watery biological environment. So the cell uses an ingenious approach. It constructs a copy of itself inside of itself, using its own cell membrane to form the protective environment for construction, and then divides by drawing the cell membrane inward between the original and the copy, eventually sealing off the gap and releasing the now completed copy into the larger environment.

This approach allows the cell to faithfully self-replicate while avoiding disastrous interfering cross-reactions with other chemicals and molecules in the environment. It also keeps the cellular components from drifting apart and being lost in the watery environment.

In a living bacterial cell, for example, the cell expands and the necessary inner components of the cell are faithfully replicated (including DNA). After the cellular components separate into separate ends of the cell, the center of the cell is divided and sealed off by a septum—a new cell wall and new membrane material separating the two halves—until it is completely sealed and the two cells are separated.

Compare this replication process to our hypothetical self-replicating 3D printer. It would be as though our printer, basically a cube-like structure, were to expand its own frame to encompass a space the size of two printers, construct and assemble the new internal components in that protected space, and then rebuild two walls between the identical sections in order to release the completed copy into the environment. This would be a remarkable engineering feat indeed!

Finally, a truly autonomous self-replicating entity must also be able to locate, acquire, and make use of raw materials for construction of new parts, and generate its own power from materials available in the environment. No convenient electrical cord plugged into the wall, please. Nor any careful feeding of the printer filament by a user. And for long-term successful replication over more than just a few generations, it would be critical to have numerous feedback and quality control mechanisms, error correction capabilities, and the like.

The above is only a very partial outline of what would be involved in building a truly self-replicating machine. But as we think through some of these details—an exercise that is, unfortunately, too often skipped by abiogenesis enthusiasts—we begin to grasp the scale of the problem. As we do so, and wonder if such an engineering feat is even possible, we should not overlook the elephant in the room: the creation of a 3D printer that could print all of its parts and also assemble all of those parts into another self-replicating 3D printer, and successfully do this generation after generation—as even the simplest self-replicating cell can do—would be a most impressive work of ingenious design.

Wanted: Molecular Unicorn, Purple

WHILE THERE are important differences between the macro world of 3D printing and the micro world of the cell, our brief review of what would be involved in creating a truly self-replicating machine in the macro world gives us a sense of the many abilities a self-replicating molecule would need to possess to be truly viable. For instance, the molecule would need to identify, position, and orient the correct individual building blocks. It would then need to link them together. In a liquid environment, the linking could not occur without accessing some energy source for assembling the blocks together. And among the many other capacities it would need to possess, the molecule would need the ability to error-check the developing duplicate molecule to prevent the essential information from degrading each time it was copied.[17]

The exercise casts into doubt the notion that a solitary molecule, however large and sophisticated, could pull off the task. For starters, it would lack a cell wall to protect itself from the violent buffeting of the surrounding chemical soup and the unavoidable interfering cross-reactions. If it somehow possessed every other needed feature of a viable self-replicating machine, it would have to be orders of magnitude more complex—more sophisticated—than any existing organic molecule. And since the whole point of proposing this purple unicorn of a molecule is to suggest a potentially plausible pathway for blind natural forces to kick-start life, then the question before us isn't whether an intelligent agent could ever assemble such a remarkably capable hypothetical molecule, but whether mindless natural forces could somehow produce it, and do so before the Darwinian process of random variation and natural selection could help, since remember, that process only begins once a self-replicating entity is in place and doing its thing.

What sort of super-molecule would be required? Neither DNA nor RNA, after all, comes even close to having the full suite of capacities needed to self-replicate all by itself. Each needs the other, and each needs the cell.

What is the cell? It's a work of nanotechnology beyond anything humans have ever built. The geneticist Michael Denton described it as a "an object of unparalleled complexity and adaptive design." Denton invites us to imagine that we greatly magnify a cell so that we could see all of the cell's components working together:

> What we would be witnessing would be an object resembling an immense automated factory, a factory larger than a city and carrying out almost as many unique functions as all the manufacturing activities of man on earth. However, it would be a factory which would have one capacity not equaled in any of our own most advanced machines, for it would be capable of replicating its entire structure within a matter of a few hours. To witness such an act... would be an awe-inspiring spectacle.[18]

Having walked through just some of the technological requirements of a truly self-replicating 3D printer, this shouldn't surprise us. When we observe that a cell can self-replicate, we can begin to ask what technologies, what capacities, would be required for that to be possible. The list is long and daunting. The sophistication of the cell is wondrous.

Researchers have attempted to identify the minimal requirements for a living cell by methodically removing proteins and seeing whether the cell continues to function. For a relatively simple parasite that depends upon its host for survival, researchers identified over 300 essential proteins.[19] Another research group studied one of the smallest and simplest free-living bacteria and estimated that twice as many proteins were required.[20] Even if we take the smaller estimate, that would still mean that over 300 different kinds of molecules are required for a relatively simple self-replicating cell, not to mention the DNA molecule and the genetic instructions it contains. The idea that a single molecule could ever perform such a task stretches the imagination.[21]

Those obsessed with the possibility of a self-replicating molecule have to ignore all of the careful analysis provided above. They have to ignore the reality that self-replication, rather than being a simple kick-starting point at the beginning of the long road of evolution, lies at the end of an extremely complicated, sophisticated, and specified engineering process. And they have to ignore the fact that every time a significant new biological feature is acquired or something is added to assist in the self-replication process, the very addition requires a reworking of the self-replication process itself, along with the likely addition of new components and additional information to self-replicate the new parts.[22] This fact, rarely discussed or acknowledged in the evolutionary literature, not only casts doubt on the materialistic origin-of-life story, but also represents a fundamental conceptual problem for the evolutionary process thereafter.

The idea that self-replication is the starting point for the origin of life is not just questionable. It is not just one more in a long list of problems with the abiogenesis story. It does not just make the odds worse. It is completely upside down and backwards. It is diametrically opposed to the physical, chemical, and engineering realities.

So why do many origin-of-life researchers remain fixated on the purple unicorn of a self-replicating molecule? Regrettably, in the area of abiogenesis it is the theory instead of the evidence that drives the thinking. This is why Richard Dawkins could confidently claim in *The Selfish Gene* that "a molecule that makes copies of itself is not as difficult to imagine as it seems at first, and it only had to arise once."[23] Yes, imagine. The theory just requires some imagination and a strong dose of luck.

It's why so much energy in origin-of-life research today is focused on finding this elusive self-replicating molecule. The insistence on a materialistic origin of life, coupled with the hypnotic allure of the supposedly limitless power of natural selection, leads the materialist to draw a conclusion that is not only unsupported, but diametrically opposed to the physical, chemical, and engineering realities we see in the world around us. That is, it flies in the face of what we know both from engineering and from observing the simplest self-replicating bacteria, each a marvel of engineering sophistication.

Self-replication, contrary to the materialist abiogenesis story, is not the *beginning* feature, a rudimentary trait that a single molecule could handle. Rather it is a *culminating* trait, one of the most dazzlingly high-tech traits in the biosphere. The accumulated evidence, taken together, strongly suggests that self-replication lies at the end of a very complicated, deeply integrated, highly sophisticated, thoughtfully planned, carefully controlled engineering process.

In the end, the abiogenesis story is not just incomplete, with details remaining to be filled in. No. The abiogenesis paradigm, with its placement of self-replication as the first stage of development, is fundamentally flawed at a conceptual level. It is opposed to both the evidence and our real-world experience and needs to be discarded.

Review: Your Turn

1. What is abiogenesis?
2. What is attractive about the idea that Darwinian evolution began with a simple self-replicating molecule, rather than a living organism?
3. According to abiogenesis researchers, is Darwinian evolution relevant to the origin of life? How?
4. How close are we to being able to create a truly autonomous, self-replicating machine? What are some challenges that still remain?
5. What problems, if any, are there with the idea of nature producing a relatively simple, self-replicating molecule?

FUEL YOUR CURIOSITY!

Recommended Resources for Further Exploration:

VIDEOS

The Basic Building Blocks & the Origin of Life

The Riddle Of Life's Beginnings feat. Biochemist James Tour

Challenge to Origin of Life: Replication

PODCASTS

The Innovative Cellular Engineering That Keeps Us Alive
Howard Glicksman

Could Blind Forces Build a Self-Replicating Molecule?
Rob Stadler

ARTICLES

Molecular Machines
Michael J. Behe

First Life Must Have Had a Minimally Reliable Replication System — A Conundrum for Materialists
Jonathan McLatchie

WEBSITES

evolutionnews.org
Original reporting and analysis about evolution, neuroscience, bioethics, intelligent design and other science-related issues, including breaking news about scientific research.

discovery.org/id/
The institutional hub for scientists, educators, and inquiring minds who think that nature supplies compelling evidence of intelligent design.

intelligentdesign.org
Documents the mounting scientific evidence for nature's intelligent design. Through this site, you can explore the evidence for intelligent design for yourself.

ENDNOTES

1. Jack Szostak, "From Telomeres to the Origins of Life," interview by Claudia Dreifus, A Conversation With, *New York Times*, October 17, 2011, https://www.nytimes.com/2011/10/18/science/18conversation.html.

2. Gerald F. Joyce, "In Lab, Clues to How Life Began," interview by Nell Greenfieldboyce, *All Things Considered* (transcript), NPR, January 8, 2009, https://www.npr.org/templates/story/story.php?storyId=99132608.

3. Despite Dawkins's optimistic suggestion that organic molecules would drift "unmolested" through the primordial soup, origin-of-life researchers now recognize that the twin problems of chemical breakdown and interfering cross-reactions pose massive difficulties for any abiogenesis scenario. Indeed, one of the key challenges for modern origin-of-life researchers is to find a way to isolate and protect the tender early molecules from the devastating effects of breakdown and interfering cross-reactions long enough for anything else interesting to happen on the road to life.

4. Richard Dawkins, *The Selfish Gene*, 30th anniversary ed. (New York: Oxford University Press, 2006), 15.

5. Dawkins, *The Selfish Gene*, 15.

6. In 1861, Max Schultze, a German microscopic anatomist, described the cell as "a blob of protoplasm, at the heart of which lies a nucleus...." Félix Dujardin, a French biologist and early pioneer in Protozoa research, referred to a "ubiquitous gelatinous substance" as a key cellular substance in common between animal and plant life. Both quoted in Mario A. Di Gregorio, *From Here to Eternity: Ernst Haeckel and Scientific Faith* (Göttingen, Germany: Vandenhoeck & Ruprecht, 2005), 67–68.

7. Charles Darwin, *The Origin of Species by Means of Natural Selection, or the Preservation of Favoured Races in the Struggle for Life* [1872], 6th ed. (New York: Mentor, 1958). Darwin makes over a half-dozen references to this "plastic" property of organisms in the *Origin*. His glossary defines "plastic" as "readily capable of change."

8. Bradley J. Fikes, "Lab-Evolved Life Gets Closer in Scripps Research Study," *The San Diego Union-Tribune*, August 15, 2016, https://www.sandiegouniontribune.com/business/biotech/sdut-rna-world-origin-life-2016aug15-story.html. The article discusses a science paper by David P. Horning and Gerald F. Joyce, "Amplification of RNA by an RNA polymerase ribozyme," *PNAS* 113, no. 35 (August 2016), https://doi.org/10.1073/pnas.1610103113.

9. See, e.g., Theodosius Dobzhansky's discussion of Gerhard Schramm's "Synthesis of Nucleosides and Polynucleotides with Metaphosphate Esters" in S. W. Fox, ed., *The Origins of Prebiological Systems and of Their Molecular Matrices*, Proceedings of a Conference Conducted at Wakulla Springs, Florida on October 27–30, 1963 (New York: Academic Press, 1965), 309–10.

10. See, for example, Natasha Paul and Gerald F. Joyce, "A Self-Replicating Ligase Ribozyme," *PNAS* 99, no. 20 (October 2002), https://doi.org/10.1073/pnas.202471099.

11. Rice University synthetic organic chemist James Tour recently reviewed some of the challenges of building organic molecules and assembling a self-replicating molecular system, in "Time Out," *Inference: International Review of Science* 4, no. 4 (July 2019), https://inference-review.com/article/time-out. See also Tour's lecture at the 2019 Dallas Science and Faith Conference, at Discovery Science, "James Tour: The Mystery of the Origin of Life," YouTube, video, 58:01, March 18, 2019, https://www.youtube.com/watch?v=zU7Lww-sBPg&t=1644s.

12. To further explore the minimal requirements of a self-replicating entity, see Arminius Mignea, "The Engineering of Life," in *Engineering and the Ultimate: An Interdisciplinary Investigation of Order and Design in Nature and Craft*, eds. Jonathan Bartlett, Dominic Halsmer, and Mark R. Hall (Broken Arrow, OK: Blyth Institute Press, 2014), Part IV.

13. See "RepRap," RepRap, September 22, 2019, accessed November 2, 2019, http://reprap.org/wiki/RepRap; see also "RepRap Project," Wikimedia Foundation, last modified January 5, 2020, 11:07, https://en.wikipedia.org/wiki/RepRap_project.

14. Jean Le Bouthillier, "BI V2.0—A Self-Replicating, High Precision 3D Printer," *Kickstarter*, Kickstarter, PBC, September 22, 2014, accessed November 2, 2019, https://www.kickstarter.com/projects/1784037324/bi-v20-a-self-replicating-high-precision-3d-printe.

15. Several other 3D printing projects have also touted their "self-replicating" capabilities, including Dollo and Snappy.

16. A search for "reprap parts" in Google Images brings up many such examples.

17. Manfred Eigen, "Self-organization of Matter and the Evolution of Biological Macromolecules," *Die Naturwissenschaften* 58, no. 10 (1971): 465–523, https://doi.org/10.1007/BF00623322.

18. Michael J. Denton, *Evolution: A Theory in Crisis* (Chevy Chase, MD: Adler & Adler, 1986), 328–29.

19. John I. Glass et al., "Essential Genes of a Minimal Bacterium," *PNAS* 103, no. 2 (2006): 425–30, https://doi.org/10.1073/pnas.0510013103.

20. Stephen J. Giovannoni et al., "Genome Streamlining in a Cosmopolitan Oceanic Bacterium," *Science* 309, no. 5738 (2005): 1242–45, https://doi.org/10.1126/science.1114057. See also a discussion of efforts to build a minimal genome by the J. Craig Venter Institute at https://www.ncbi.nlm.nih.gov/pmc/articles/PMC4879981/.

21. Consider a minimal cell model discussed by In Vivo Veritas at http://invivoveritasest.blogspot.com/2013/07/a-minimum-cell-model-and-origin-of-life_4.html.

22. This need not be an infinite regress. Based on what we see in biology, the self-replication process apparently can be engineered and the challenge can be overcome. But we do start to sense the scale of the problem.

23. Dawkins, *The Selfish Gene*, 15.

IMAGE CREDITS

Figure 1. 3D printed cube stand. Photographs by Eric H. Anderson.

Figure 2. RepRap printer. Image by RepRap Project, 2007, Wikimedia Commons. CC BY-SA 3.0 license. Descriptive arrows added.

WHAT IS THE DISCOVERY SOCIETY?

The Discovery Society is a group of individuals who come together to support the work–and disseminate the message–of Discovery Institute's Center for Science and Culture. New members receive materials that help educate themselves and spread the word about our work to those in their circle of influence. Depending upon their giving level, members receive one to three Discovery Institute Press newly released books per year, along with invitations to regional donor events and discounted rates on our annual Insiders Briefing events.

If you appreciate this booklet and aren't already a member, we hope you will consider joining our network of supporters today!

Your donation to Discovery Institute's Center for Science and Culture will allow us to expand our cutting-edge scientific research and scholarship; train young people through our education and outreach; and reach the masses through media and communications.

discovery.org/id/donate

MORE INFORMATION ON THE DISCOVERY SOCIETY CAN BE FOUND AT
discovery.org/id/donate/#member-levels.

EVOLUTION AND INTELLIGENT DESIGN IN A NUTSHELL

Are life and the universe a mindless accident—the blind outworking of laws governing cosmic, chemical, and biological evolution? That's the official story many of us were taught somewhere along the way. But what does the science actually say? Drawing on recent discoveries in astronomy, cosmology, chemistry, biology, and paleontology, *Evolution and Intelligent Design in a Nutshell* shows how the latest scientific evidence suggests a very different story.

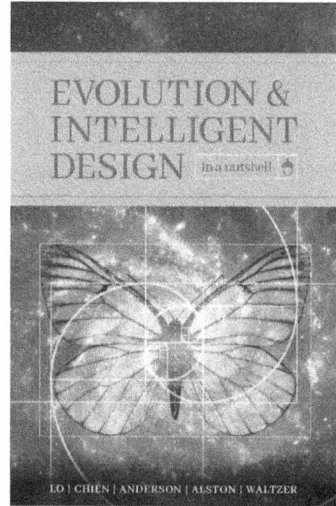

"accessible, informative... powerful ... an excellent resource."

J. Warner Wallace

PURCHASE THE FULL BOOK HERE:

DiscoveryInstitutePress.com/EvolutionandID

MORE IN THIS SERIES:

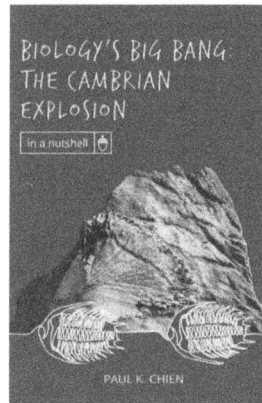

THE BIG BANG &
THE FINE-TUNED
UNIVERSE
in a nutshell

ROBERT A. ALSTON

THE ORIGIN OF LIFE
& THE INFORMATION
PROBLEM
in a nutshell

ERIC H. ANDERSON

EVOLUTION'S
IRREDUCIBLE
COMPLEXITY PROBLEM
in a nutshell

ROBERT P. WALTZER

BIOLOGY'S BIG BANG:
THE CAMBRIAN
EXPLOSION
in a nutshell

PAUL K. CHIEN

This series of booklets was created to help Discovery Society members educate themselves about the basic arguments for intelligent design and the critiques of Darwinian evolution. Each booklet presents the content of one chapter of *Evolution and Intelligent Design in a Nutshell*. To help you delve deeper into each subject, we have included a list of recommended resources from our vast library of videos, podcasts, articles, and websites. Members of the Discovery Society can download digital versions of these books through the Discovery Society Community on the DiscoveryU.org platform or purchase physical copies at a discounted rate through Amazon.com.